THE ENDLESS UNIVERSE TIME & GODS

SANDEEP BISHT

"The Endless Universe Time And Gods"

Contents

Contents

FOREWORD

Praise from readers around the world

<< *Great book !!! Well researched with amazing coverage of subject, and presented in intresting absorbing writing. An ideal candidate to become a textbook for secondary grades. Looking forward for another book, Cheers >> USA*

<<Having been fascinated with world history, I found this book very illuminating. In this relatively short book, an engaging time-travel story takes one through voyages in time and place, chapter after chapter, while unfolding and connecting so much forgotten events that have influenced humanity since the cataclysms that ended the last ice-age and created the 'Flood' until today. A MUST READ FOR EVERYONE THAT WANTS TO MAKE SENSE OF TODAY'S WORLD CONDITIONS. >> Canada

<< I really enjoyed this book, and was immediately drawn in by the writer's creative and engaging invitation to take a journey through time and space and to discover historical events and symbolism that have significance in our current lives. This fact-filled novel tickles the reader to think, research, and ask some profound questions. The stories are written beautifully so I found myself sharing excerpts with friends and family. I have recently read that what makes a man a great man is the pursuit of questions. This you have shown in your writing, and inspired in your readers. Thank you Siamak for sharing your work, adventures, and thoughts with us. >> Ukraine

PREFACE

ACKNOWLEDGEMENTS

PROLOGUE

I

The Endless Universe
Time & Gods

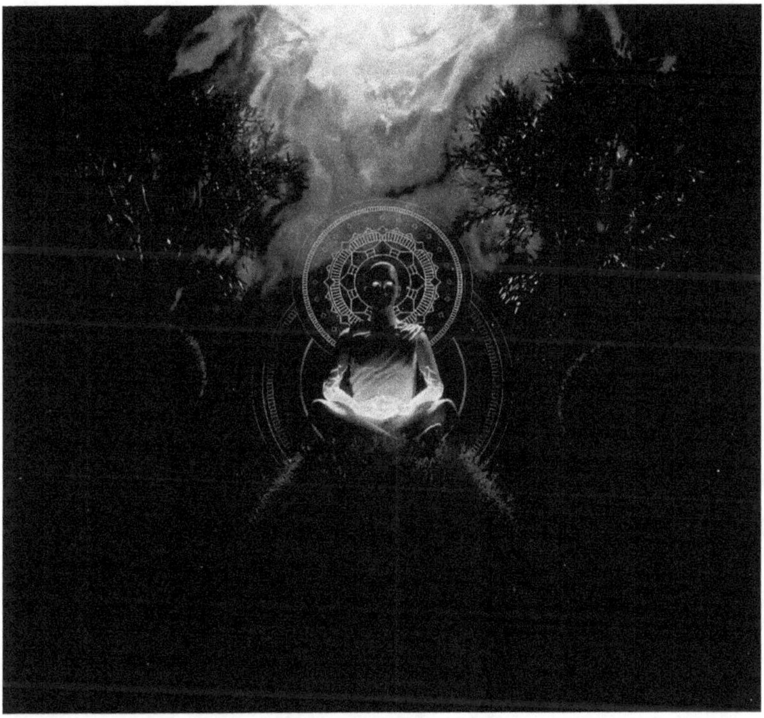

THE FIRST PRINCIPLE

Humanity seems to emerge into existence and a world, unkowing and destined to leave still to a great degree, unknowing. Much of what one encounters, from the nature of fellow humans to the structure of existence itself, serves as a cloche and barricade to ultimate reality. The smallest fraction among them will cast aside the human proclivity for fantasy and delusion, find and cling tenaciously to a single absolute. Armed with nothing but this one absolute, they assault the barriers to knowledge and understanding. They map have no academic, intellectual, or scientific status or even any allies in their quest, but it is they who are the true philosophers.

While it is said to be difficult to define, metaphysics is simply the study of existence, as such. It is a search for truths and a comprehensive and fundamental understanding of existence. The principle of non contradiction, for example, is a principle of being as a such. It is germance to the study of economics, but it is not a principle of economics per se; it is relevant to the study of physics, but it is not a law of physics per se. The law of non contradiction, which states the law of identity in reserve, is a law of metaphysics, of existence as such, and is pertinent to everything.

Metaphysical study should begin with an examination of the axiom and ask the question: "What may one hold to be true of reality, by virtue of the fact one knows that all A is A?" Obviously, metaphysical investigation should not require special knowledge, limited to a certain field and, in fact, should not constitue specialized knowledge, as it is general and applicable to all study. Today, Unfortunately, metaphysical examination and pondering is most certainly and exceptional effort. Metaphysical investigation itself has become heterodox. It is entirely out of fashion.

Instead of rational metaphysical inquiry, there is metaphysical deconstructionism serving only the purposes of making meaningful philosophical throught impossible. While it deceives the practitioner into believing they are winning a debate over reason itself, it is only their own minds that they are negating. Philosophy has degenerated into the art of obscurantism. It has become a discipline worthy only of the attention of confidence men, shyster lawyers, politicians, and other professional liars. This does not refute the meaningfulness and importance of a true metaphysical inquiry and discipline.

True metaphysical investigation must begin with the axiom. It is the fountainhead of all human knowledge. It cannot be replaced with empirical science.Nor can it be replaced with religion, which is just primitive, arbitary and rationally unjustified metaphysics. The sad state of this noble and monumentally important pursuit of truth is exemplified by the fact that the very word "metaphysical" is now often seen as synonymous with a belief in magic. Metaphysical questions are associated with mental illness. Perhaps seen as a threat to religious fantasy, abstinence from serious metaphysical inquiry is practiced with religious devotion.

While it may be primarily used as an alternative term for metaphysics, the two fundamental branches of philosophy, metaphysics, and epistemology, the study of the means by which we acquire knowledge, may be combined and referred to as first philosophy. Metaphysical matters are inescapable. The attempt to separate epistemology from metaphysics is foolish with a predictable nihilist result because, in the end, it divorces epistemology, and therefore the very pursuit of knowledge, from reality. It is also significant to note that when the metaphysical philosopher embraces the axiom as the supreme law of existence, this constitutes a fundamental epistemological claim to knowledge. The absolute ground of existence and knowledge are discovered together, and the fact that one is discovering both must be recognized. It is not enough for the axiom to be regarded as a principle of reason. It must be recognized as metaphysical truth; it must be acknowledged as ontological or it is not meaningful as a principle of reason.

So, these branches of philosophy, metaphysics, and epistemology, in fact, must be approached together, as they are intimately related, and it would be impossible successfully to address one separate from the other. They are, in fact, founded on the same first principle. The first principle of reason is also the supreme law of existence. It could not serve as the first, without being acknowledged as the latter and the axiom could not be known as the supreme law of existence without being acknowledge as fundamental knowledge. The law of identity, standing and regarded as just a principle of reason has had a calamitous effect on philosophy and the human mind. Knowledge is grounded in metaphysical principle, and it must be regarded as such. The absence and rejection of metaphysics make knowledge impossible.

....

It is an incontrovertible truth that all A is A. Things are what they are. Everything else that exists must possess self-sameness. This, the law of identity is the most basic axiom. An axiom is a self-evident truth; it proves itself. Axioms are also sometimes referred to as necessary truths because it is rationally inconceivable that they could be false.

There is an incalculable number of necessary truths, but there is only one basic axiom, the law of identity. Many common sense assumptions has contributed to the erosion of man's intellectual confidence while confidence while conversely, his worldview ha sbecome more sophisticated. The axiom is not an assumption, but when widely held common sense assumptions turn out to be false, as they often have, the axiom is seen to have fallen.

Self-evidence declares identity. Any axiom, if indeed it is a true axiom, is such because it asserts self-sameness. This is the monistic view of axioms. The law of non-contradiction simply states the law of identity in the negative. It states there can be no non-identity. The most basic example of a contradiction may be expressed as A is not A. It by no means trivial to point out that this also the most basic example of a lie.

Such a blatant lie seldom stands naked, but it is the vulgar fraud that steals beneath the glorifieed loose-mouthed insinuaton of mysticism. It is, therefore, poetic irony when the critics of non-contradiction fall back on a childish puzzle fittingly called the liar's paradox. If one wishes to understand the absolute corruption of the mysticism that grips humanity, it is found in the recognition that thev embracing of condradiction is not only a lie, but that it is the fundamental falsehood that all lies mirror.

It was with profound irony when the British philosopher Bertrand Russel lamented "one of the painful things about our time is that those who feel certainty are stupid, and those with any imagination and understanding are filled with doubt and indecision. "This same

philosopher once reffered to philosophy as "on the whole a rather hopeless business." Knowledge, justifiable certainty, is undoubtedly not discovered as easily as the false confidence of the fool. Humanity, nonetheless, need not be doomed to observe mindless fools follow mindless fanatics as the more intellectually talented among them stand idle and endlessly confused, This is the result of the failure of philosophy as Bertrand Russel and others seem to be implying, but never explicitly accepting blame. It is philosophy, nonetheless, that has brought this condition to humankind, and philosophers such as Russel have done little to change this. It is the failure of the philosophers,
the result and perhaps the purpose of such philosophy that has turned any claim of certainty into the delusion of the feebie minded.Once comprehended and understood, nonetheless, the real claim to, and ground of knowledge is an idea that could be grasped even by the less intellectually gifted. The self-imposed thoughtlessness and entrenched stupidity that has dominated humankind has been the result of self-deception, much more than a lack of intelligence.

It is the function of philosophy to formally and explicitly formulate the axiom and recognize its significance and meaning. Historically, philosophy has failed and humanity is offered mindless indulgence and mystic fantasy as the only alternatives to unfulfilled, intellectual effectiveness and potency. This has constituted the most appalling and catastrophic failure in human history and experience.

At least, nonetheless, a latent, implicit acknowledgment is and must be present for humans to think. Necessary truth is implicit in any rational assessment of perceptual information. This implicit common sense has never been completely transformed into knowledge. Unable to achieve its proper status, this common sense, loathed and attacked, can only depreciate and decline. This can be seen in the horrendous intellectual environment, especially in a social or political atmosphere or controversy.

Even among those who present well-reasoned ideas and arguments, when viewed and scrutinized in the broader context of the philosophy or religiosity they embrace, would make such reason groundless and unprovable. Reason is embraced and employed in some matters while other subjects are left to fantasy and mysticism, seemingly unaware that one negates the other. For one who understands this, the abject absurdity of this compartmentalization is truly disheartening. It is not a matter of embracing the right beliefs. for it is not right to just believe. Humanity must rise above belief and discover knowledge.

Objectivity is rooted in the fact that the axiom is the most supreme law of existence. The first principle can provide us with a simple, provable concept of objective reality. The objectivity of reality simply means that things, primarily, are what they are. The objectivity of reality is grounded in the fact of identity. It is significant to note in this context that clls to accept objective reality in hard to accept circumstances, are often expressed with an axiom such as "well it is what it is."

Even if one assumed that one is in a world such as what portrayed in the fantasy film, The Matrix, a world where all perception is somehow manipulated would not obliterate our basic metaphysical concepts. It would still have a reference to something that exists, thought something very different than what we thought. Existence nevertheless,

would still be objective. If one lived in such a world, then one lives there, and that would be an objective truth. As mystics have stated with contempt, identity and the law that asserts it is immutable. This is what objective reality means within the context of the philosophy of identism, the philosophy presented in this work, and it is the only conception of it that it embraces. Things do not exist independent of thought and feeling. Thought and feeling exist; they exist as factors. Even the existence of subjectivity is objective. One's thoughts, hopes, or prejudices, while they may not represent the exercise of objective reasoning, are part of reality and possess identity. Whatever its causes, and effects they may have, the existence of subjectivity can only be so because its existence concurs with the fact of identity.

The formulation of the axiom is the simplest but most profound product of human creativity. That simple act of creativity provides the most basic premise of reason. It is this modest unassuming but certain truth that elevates man's notions about the world to the lofty status of knowledge (i.e, truth held with justifiable certainty.) All knowledge, even firsthand perception, is ultimately validated by this first principle and the existence of identity.

Logic and perception are both founded on the same first principle of metaphysics. Logic might be characterized as applied metaphysics, but much to the discontent of empiricist, so too, could science and the scientific method. Axioms are basic, straight on, assertion of identity, then it is not a logical truth. greatest invention in human history. Without a medium of exchange, advanced civilization would be impossible, but without at least an implicit embrace of the axiom, of logical truth, no human reason is possible. Any evalutaion of perceptual infomation involves the employment of logical truth, interestingly, historically, necessary truth is perhaps the only thing that has been despired, scorned and ridiculed more than money. Criminals and mystics have managed to take both of these greatest invention in human history. Without a medium of exchange, advanced civilization would be impossible, but without at least an implicit embrace of the axiom, of logical truth, no human reason is possible. Any evalution

of perceptual infomation involves the employment of logical truth, Interestingly.

historically, necessary truth is perphaps the only thing that has been despired, scorned and ridiculed more than money. Criminals and mystics have managed to take both of these greatest of human values and twist, pervert and manipulate them into weapons of control and plunder. Nonetheless, the concept of identity has been embattled by the mistake of the well intending. Such is the case with those who have held identity to be synonymous with existence.

....

It has been maintained that to hold identify as a part of reality is much like seeing identify as a coat of paint applied over a house. This is not a good analogy at all for the existence of self-sameness. Identify is not an afterthought. The notion that the identist concept of identity is platonic, Implying some sort of abstract world is false. The concept of identity is abstract only in the same respect that all parts of existence are abstract. They are parts mentally abstracted of the totality of existence. The platonic notion of some domain of a ghostly abstract outline of reality is particularly inapplicable to identity. There is nothing that can be abstracted from identity. Identity is irreducible. Identity exists in the only reality that it creates.

Identity is not a product of the things that possess it. The things that possess identity, everything else that exists, is ultimately a product of identity. A much better analogy than a coat of paint would be to compare identity to the atoms that comprise the house. One can strip away the paint; yet, one would still have a house. On the other hand, you could remove paint, doors, or windows. Whatever parts you remove, the house would still contain atoms and so too, would the parts that you remove. Self-sameness is even more fundamental than atoms; its removal is rationally inconceivable. The law of identity does not grant us omniscience, but it does tell us something that is true of everything. If we fail to acknowledge this, then we know nothing. Consider such a phony esoteric statement as "the law of identity breaks down at the quantum level" or "the subject transcends the native of human reason". One could use lost socks in the washing machine as evidence of an unknowable world of contradiction, and it would be less sophisticated, but no less idiotic. This is arriving at a contradiction in one's thinking, and then blaming reality. This is the ultimate intellectual dishonestly and corruption hiding behind academic status. This is baseless arrogance more like a spoiled child than a scientist or intellectual.

The critical importance of the axiom's relationship to knowledge is stated in metaphoric eloquence with the expression "the buck stops here" However impeccable or imperfect. It does not matter if the information we acquire about reality and our world comes from the daily newspaper, the internet, firsthand perception or a little voice in one's head coming from little green men from outer space. All our knowledge and even the very concept of truth itself ultimately rest on the axiom. It is not necessary to know everything about the means by which we acquire knowledge to prove this. All that is necessary is the recognition of the immutability of the axiom and the existence of self-sameness.

To constitute an unwavering foundation for knowledge or even justified opinion, however, the axiom must be uncompromisingly acknowledged as an all inclusive immutable absolutte. The ultimate implication of just one adulterous with absurdity is the abdication of any claim to knowledge. Even the concept of truth itself is obliterated when the identity principle is betrayed. If one imagined a fantasy domain of non-identity, logic and mathematics would be nothing but mind games with completely arbitrary rules but also seeing would by no means justify believing. When embracing the delusion of embellished contradiction, often peddled as a limitless possibility, one possibility that must be surrendered is knowledge.

Knowledge begins when the axiom is formulated, acknowledged as incontrovertible, but also, it must be recognized that it asserts the existence of identity. If one doezs not acknowledge that all A is A, then one has relinquished the claim of any knowledge. Even firsthand perception becomes ambiguous when the axiom is not embraced with absolute certainly. Furthermore, it is not just knowledge in the absolute sense that will fall. Any claim of likelihood will also sink into this bog of uncertainly.

The best of Western culture brought to the world the ideals of reason, individualism and intellectual and economic freedom. but they were never fully realized and correctly defended. Because of this failure and the natural hatred of these are devious and powerful criminals working to eradicate what remains of Western culture from the planet. This could be the price of western philosophy's failure.

Regardless of the fact that influence of reason has been relatively brief and latent weighed against the domination of mysticism, the embracing of contradiction, its effects have been profound. While the understanding was flawed, the foundation of man's greatest achievements and the accomplishments of Western culture is the first principle and identity philosophy. Philosophy that acknowledges the axiom as absolute truth and profoundly important. It is no coincidence that one preceded the other in history. This causative relation was demonstrated twice, first in ancient Greece and then in Europe. fostering the age of reason. Islam also had a brief enlightenment as it was influenced by Aristole and other Greek philosophers. Apparently, explicit identity philosophy emerged from implicit common sense that was temporarily liberated from religious Superstitions.

Humanity, However has never really discovered knowledge, for the validity of one's claim to any knowledge or even justified opinion is ultimately lost when the axiom is betrayed. All opinions become equally arbitrary and baseless. If one decides the identity principle is not true in some mystical domain, then one loses the justification for assuming self-sameness exists anywhere. This is not a hypothetical implication of falling to acknowledge that all A is A. It is a precise description of the collective state of the human mind. The originality of this writing is evidence that knowledge and comprehension of identity are not inborn. To the contrary, humanity has struggled to understand necessary truth, its meaning

, and implications.

All A is A Only a corrupt mind that is divorced from reason will fail to acknowledge that the law of identity must be universally true and all-embracing. This fact is not a meaningless truth. Philosophy's greatest historical tragedy is the failure to recognize that the axiom is ontological.

Assertions are true in the respect that they agree with the part of reality to which they refer. Meaning, reference to reality is a prerequisite for truth. The notion that the assertion "A is A" does not state anything about "A" is shallow and false. It does not assert anything about "A" that distiguishes it from anything else because what it declares of "A" is true of all things. The identity; self-sameness is the existing thing to which the axiom refers. It is incontrovertibly true that a dragon is a dragon, but yet, there are no dragons. What's then is the truth of the statement ? The statement "dragon is dragon" does not assert and prove the existence of dragon; it asserts and proves the existence of self-sameness. Logical truths are complex assertions of identity. The mathematical statement "2 and 2 are 4" may be reduced to "1 and 1 and 1 and 1 are 1 and 1 and 1 and 1", and this asserts a truth even if you are counting dragons. The truth that it asserts is the truth of the existence of identity. Logic asserts identity. The statements 2 and 2 are 4, A is A. All logical truths and all axioms reference the same fact, the existence of identity. Their application is diverse and universal because the thing to which they refer is the essence of everything. Alllogical truths are complex assertions of self-sameness. It is the fact of the existence of identity that grounds logic to reality.

The law of identity is the supreme law of existence, but it does not govern the universe. It is the existent thing to which that law refers, identity, which rules the universe. Contrary to assertions such as "laws govern the universe." laws govern nothing. The apple doesn't fall from the tree because of the law of gravitation; it falls because of the existence of gravitation. Newton's law of universal gravitation is a description of something that exists; It describes gravitation. Whether it is warped space or something else, his law, nonetheless, refers to something that exists and so too does the law of identity. But while Newton's theory may be flawed description of something that is complex and conditional, the law of identity is a perfect description of something that is absolutetly simple and unconditional.

Once the contradiction of meaningless truth is exposed with, it becomes clear that the identity principle proves the existence of identity. The fact of identity's existence is the fact that hinges reason to reality. The reult of not acknowledging this fact is the detachment of logical truth from reality and rendering reason mute without ever having to deny it's "truth".

Truth itself follows the axiom into irrelevance. This is mysticism's most devious, insidious and corrupt achievement. The failure to acknowledge that the axiom is ontological, not just a principle of reason, is catastrophic because its acknowledgment is indispensable in its role as the foundation of human knowledge. All axioms, all logical truths, ultimately reference the same monistic fact, the fact of the existence of identity.

The denial of the existence of self-sameness leaves humanity hopeless uncertain. The postmodernist nihilist mystic seems to be fully aware of this when reason is described as "the weighing of notions against imaginary ideas." This is an attack on human knowledge of deadly sophistication, aimed at the very heart of reason. Most certainly, nonetheless, reason is not the weighing of notions against imaginary ideas. It is the weighing of ideas against the incontrovertible and universal fact that the mystic dreads, the existence of identity. The existence of self-sameness and the ultimate reality it implies is what there delusions and fantasies seek to escape.

It is axiom itself that proves the existence of identity. This is the ultimate reality it implies is what there delusions and fantasies seek to escape.

It is the axiom itself that proves the existence of identity. This is the essential recognition that has been lacking. It is this fundamental absence that has detached the human mind from reality. It is this wanting fundamental knowledge that could unify metaphysics and epistemology and thus, reality, to the mind. The axiom provides the most basic knowledge of existence, the existence of identity. No true knowledge is possible without this foundation.

One may know nothing else about what is on other side of the universe, but one can know with absolute certainty that whatever is out there, that is what is out there. It must possess self-sameness. In this respect knowledge is primarily a priori. Yes, even empirical science in this fundamental regards is a priori. Empirically derived knowledge must be founded on that which can be known, and only be known a priori, the existence of self-sameness.

The validity of perception and the fundamental truth of reason are both founded on the first principle and what it asserts, the existence of identity. Only identity can serve as the foundation of knowledge. Any other claim of an underpinning is a fraud and any notion that knowledge does not require such a foundation is equally fasle. When the axiom is betrayed fundamentally , choosing consistency over inconsistency becomes an arbitrary choice. Without the underpinning of the existence of identity, logic itself will often be seen as an arbitrary choice that is inconvenient and imposing.

Empiricism cannot save the human mind from the failures of rationalism. Science and humankind cannot run from the imposing metaphysical questions, with all their social, ethical and political implications. Such issues cannot be successfully addressed if a basic understanding of existence and the essential basis of knowledge is not comprehended. Empirical science, rational metaphysics, human freedom and civilization with live or die together.

What social, ethical and political system is possible and appropriate for the mindless, the blind and the deluded? Such a humanity is just sheep for the control and slaughter of whatever criminal gang can prevail. Common sense cannot escape the world and the universe.

The law of identity is the simplest of human principles, but it's true and profound meaning remains latent and undiscovered. Within this vaccum, it is the claim to human knowledge that remains concealed with this undiscovered absolute. For this absolute is the fountain head of all else.

THE MONUMENTAL DENIAL

Proof is that which supports the truth of what one is asserting. Proof in the absolute sense is that which proves absolutely, that what one is asserting must be true because any alternative is impossible. There may be relatively little one can know with this level of certainty, but the existence of identity most certainly can be held as immutable and, as it happens, must be. Without thsi recognition, there is only the prospect of intellectual chaos and confusion. This is simple, basic, and incontrovertible. Not with standing humanity's accomplishments, the grim state of humankind, human culture, and humanity's tenuous and ultimately fraudulent claim to knowledge is a powerful and disturbing demonstration of the ultimate meaning and implication of this failure As this failure As this failure demonstrates, this rationally indisputable knowledge is not innate. It comes from the axiom and a recognition of its meaning, the existence of identity. Once confronted with the fact of the existence of identity, it is only by virtue of the human mind's capacity for self-deception that identity could ever be denied, and usually, this denial is implicit, not unambiguous. "Denier" has become a demonized term, but surely the failure to recognize and acknowledge the immutability of the identity principle and what it asserts is the ultimate irrational denial. In contradistinction to the historical mistake of rationalism, however, this knowledge is not the product of some inner light or intrinsic knowledge. It is the knowledge gained from the axiom itself; it is the axiom that asserts and proves the existence of identity. No set of empirical facts can ever extrinsically prove that which is proclaimed by the axiom. Without the fat of identity, nonetheless, there will never be confidence and justification for calling anything a fact. Contrary to the claim that the axiom is somehow supported by extrinisic facts, that we know that A is A, because we observe the structure of reality, the proof of the axiom is intrinsic. Trying to prove the axiom extrinsically puts the cart hopelessly in front of the horse. If it is not acknowledge as self- proving, then the identity principle can never be extrinsically proven. Such is the state of the human mind.

Even the truth of that which is self-evident, the very truth of self-evidence, grounded in the axiom. Without the axiom, even self-evident perceptual evidence of all else would not constitue knowledege. Hypothetically speaking, not even empirical omniscience could extrinsically prove the axiom. If one had empirical evidence of everything, save identity, the recognition and ascendance of this information to the status of knowledge would still require the fundamental awareness of the immutability of the axiom and the existence of identity.

Perception is not the thing perceived. When dealing with the mystic's claim of some alternate form of knowledge or truth, it is important to make this distinction. If a tree falls when there is no one around, then there should be the presence of sound waves. Sound, however, as an element of consciousness would not be present if there is no one to hear. This is not negated by mind, physics in differentism, the view that what we subjectively know as consciousness is the sum of the physics that constitutes it. The physics that constitute sound waves would be there but the physics that constitute hearing would not.

However we may obtain information, no information we may acquire about reality is exempt from the first principle. To reveal a fallacy or misconception as such, it is often necessary to examine it in a broader context. To do this, however, one must first acknowledge that existence does not contradict itself, that it possesses self-sameness. Mysticism is an attempt to escape this scrutiny.

Mysticism is a revolt against the law and concept of identity, a revolt against reality. Mysticism is the antithesis of identity philosophy and, as such, is driven to destroy its opposite. Sinking to the bottom of their sewer, beneath the embellished variations, mysticism is the embracing of contradiction, of non-identity. In an epistemological context, it refers to the notion that knowledge can be founded on something other than identity. Mysticism is a belief in non-identity.

Nihilism is a kind of mysticism, for to reject the self-evident fact of perception, or at least sensation, constitutes a rejection of identity and the embracing of non-identity.

••••

Whatever diverging and often violently competing, fantasies embraced the metaphysics of mysticism is the notion that existence is contradictory and irrational, that it does not corresponding with the first principle. Abdicating any claim to true knowledge, mysticism must choose between nihilism or pretentious, conjured alternative to knowledge. This fabricated claim of higher intuition feeds upon the limitations of human understanding and self-awareness. Exemplifying the often-latent workings of the human mind, once a man was resting in bed by his wife watching television. There was an attractive young woman on the television show they were watching. She seemed familiar to him, but he couldn't place her. He could not remember whom she reminded him. His wife, also watching, made the remark that for some reason she did not like this young lady, but she did not know why. She said she was certain it was

not jealousy, but there was something about her she just didn't like. Well, about that time the man realized who it was this young lady reminded him of. It was his mother-in-law. She didn't look like her; she just had a manner about her that was similar. The man, having some measures of commonsense and wisdom, never told his wife of whom this young lady reminded him.

As illustrated by this true story, the workings of the human mind are often concealed and not completely understood. One may have hunches or insights while not being explicit aware of the reasons. In this context, it is important to note mystics do not hold a legaltimate monopoly on this intuitive mode. This kind of insight is not proof or even genuine evidence of some mystic awareness outside the realm of reason or perception. If one has a feeling or intuition, it is imperative to acknowledge it as such. A rightful seeker of truth must always strive for the intellectual rigor necessary to uncover the hidden reasons behind such insight. It is by this means that one may discover whether a hunch or insight has any claim to legitimacy or rightfulness. Without such rigor, all one really has is unsubtantiated prejudice.

It is critical to our claim to knowledge that we understand the axiom, in and of itself, tells us about existence. Identity is all it asserts and all that it accounts for. Just as necessary truth is monistic, so too is a priori knowledge. It tells us only one thing about reality, but what it tells us is abosultely critical. Despite the monistic nature of this knowledge, its application is enormous. It is so tremendous that it is easy to be unaware that it is ultimately just one fact that is being referenced, knowledge without experience, a priori, is monistic and has remained latent in the mind of humankind. Aside from the existence of identtity, our awareness and understanding of the universe are ultimately derived from perception. Perception and the human mind are not infallible. A delusional mystic may really hear a voice in his head, but this does not mean he is really hearing from god or little green men. The presence of feeling, sight, sound, etc. is self-evident proof of their existence.

Perception is sensation acknowledged as information about reality. Sensation is, to the one who senses, self-evident proof of the existence of sensation. The existence of sensation is proof that something exists even if one assumes that sensation is all that exists. Common sense assumes this surely even in the mind of a child or an animal. However, this proof implies acknowledgment of the axiom and would vanish without the recognition of identity. This is exactly what has happened to the mind of humanity.

In an atmosphere of philosophically groundless chaos, relativism has emerged from the nihilism. Relativism accepts the nihilistic notion that true knowledge is impossible and attempts to replace it with some kind of intellectual elitism. Relativism is a sort of fiat reason only unlike fiat currency that is forced upon humanity by government and financial elite, to replace real money, fiat reason is offered as a replacement for knowledge forced upon us by bankrupt philosophy and intellectual establishment. Just as the result and design of fiat currency is manipulation and theft, so too is the result of fiat reason the manipulation and theft of intellectual potency. One establishes a false and criminal financial elite, the other a false and deceitful intellectual elite.

While relativism, which is just soft nihilism, is sold as a counter to extermism and dogmatism, the ascending of the false or unproven to the status of knowledge, it is actually the most intellectually crippling dogma of all. One may hold the notion that the world is flat, and it could be a popular and accepted view. The proponents of such an opinion may have academic status and prestige, but as long as an objective standard of proof is in place is in place and acknowledged, such a view can be refuted.

Relativism however, undermines any objective standard for proof and turns the battle of ideas into a popularity contest. New ideas are seldom popular and the truth even less so. Since real knowledge is regarded as impossible, only social convention controlled by elites, new ideas need not be refuted or even earnestly examined. Ideas that threaten their crippling status quo can be swept away by simply branding them as "out of the mainstream."

Ascending from an environment of relativism and intellectual elitism acknowledging the existence of self-sameness and the immutability of the law of identity would truly be the ultimate triumph of the rule of law in the most fundamental sense, over the rule of men, and their fiat reason. This would be precondition for such an achievement in the political realm.

....

Just as fiat currency appears doomed, so too, does fiat reason, Relativism appears hopelessly condemned to fall in the catastrophe it summons. It is no contradiction that while stagnating relativism dominates the intellectual establishment, it has been accompanied by head-cleaving religious fanatics who represnt the extreme of the fastest growing religion in the world. Where will the "buck stop" when the promissory notes of fiat reason must "float" against the currencies of irrationality, the basest of human emotions, in the blood-drenched financial market of history? Relativism has left the wreckage of Western culture so intellectually bankrupted that against the most irrational ideologies and philosophies they can only fall back on slander and lies as a weapon and protection. The same degenerate philosophy that so stubbornly and stupidly rejects any possibility of rational certainty, so easily falls to the mindlessly faithful, so long as that faith comes from outside the culture that still holds a tenuous
grip on the value that they most abhor.

A violent, aggressive Christianity swept through Europe replacing indigenous religions, and now a more passive, peaceful Christianity is under attack from a more aggressive, savage, brutal, inhumane Islam. There is a reason why the so-called "great religions" have such a violent history. It is why they become the "great religions."

If one were to invent some fantasy religion, but kept these beliefs to oneself, then that religion would die with its creator. Religions live or die by the law of the jungle as it is asserted by natural selection. Religions must survive and procreate just as living organisms. If not, they are lost and forgotten. Intellectual matters can be settled with facts and reason, but history has shown that faith is best promoted with the sword. Having faith in peace, compassion, and tolerance leaves one disarmed against those who have quite a different sort of faith.

Beliefs, groundless claims to truth, are all equally arbitrary. Faith, in itself, constitutes a violation of the absolute. It involves the contradiction, "unknown is known." If one decided to believe there is life on Mars without the justification of supporting evidence, this would remain a contradiction and, as such, an untruth even if it happened that there was actually life on Mars.

Life on Mars, or not, it would be false claim to knowledge. Guility or innocent, a man would receive an unfair trial if he was convicted by prejudice rather than evidence. The failure to uncompromisingly acknowledge the axiom as an absolute is to lose all claim to it. This is why faith and belief sabotage all claim to any knowledge. Knowledge and faith are incompatible. The nihilist relativist mystics of hypocrisy and disbeilf have no answer for violent believers of non-identity, for this too, non-identity,is the belief of the two-faced relativist. Metaphysical issues are far more important than most would think. Humanity's quest for its metaphysical footing may be a matter of survival. It is time to look beyond the false elite's manufactured eclips "terrorists," however are just a tiny fraction of the forces of brutality and irrationality that are arrayed against humanity. The seldom acknowledged, but scarcely deniable fact, is that most of the terror in the world is perpetrate by the gangs of criminals

called government. They are the real threat to the very survival of the human race. Behind the embellished incantations ad pretentious formality, government is nothing but the most dangerous form of organized criminality and statism is little more than the most virulent form of criminal insanity. One cannot formulate a better definition of criminal insanity is little more than the most virulent form of criminal insanity. One cannot formulate a better definition of criminal insanity that the willingness to engage in criminality while not only feeling justified but even virtuous. Underneath the varying ideological differences, this is the essence of statism. The state is nothing but the delusion of authority to commit crime. The power of this delusion is derived from the fact that most of its victims share the same delusion. It is they who must be liberated from their own delusion.

This liberation can only come about with an uncompromising commitment to reason, the first principle and the existence of identity. It must be based on that which virtually all human achievement has been grounded. Humanity must discover the shining beacon of knowledge before it can discover life and happiness giving freedom.

Neither the delusional senselessness of mysticism nor the criminal instanity of statism have really done anything positive to promote the ascent of humanity. To the contrary, they have always stood in resistance to it. As humankind has advanced, however, they have both become more dangerous, more incompatible. In the case of statism, one need only look at the trend of history to see its escalating danger, and the fantasy of mysticism can provide no protection. To the contrary, it enables it, for the state itself is nothing but a mystic hallucination.

There is one hope for the future of humanity. Mysticism and statism have risen together to dominate the mind of man, and if identity can emerge as the acknowledged absolute, they will fall together, as well. it is the periods in history in which humans came the closest to discovering knowledge that they also came the closest to liberating humanity from these monsters. If humanity does discover knowledge, with its empowering intellectual confidence, governement, and religion, the two scourges of humanity should soon be thrown together on the scrapheap of history and cultural evolution.

Meanwhile, in this frightrningly paradoxical age of confusion, mysticism, statism and nuclear weapons, a call for a gentler faith and misguided intellectual responsibility, but there is no evading the consequences of failure. Reality and identity are most assuredly, deniable, but just undoubtedly, can never be defied.

WHAT IS GOD OF UNIVERSE

The universe is apparently orderly and seems to correspond to univesal laws has been sighted and claimed as evidence for a creator god. Indeed, it is and must be an orderly universe, and this necessity is revealed when one acknowledges what the God of the universe must be. It is an achievement that can only come from metaphysical philosophy and an understanding of the axiom. Disorder may be a meaningful description of an adolescent's room, but in a metaphysical context, it is an atrocious concept. Existence is and must be, fundamentally rational because the first principle of reason is absolute.

One need not hope for order in the universe; one needs only to hope that one can find order. This does not constitute faith in reason, which would actually be a contradiction and subversion of reason. Faith, of any kind, is a violation of reason's first principle. The acknowledgment that existence is non-contradictory is not faith. Rather, it is the recognition that the metaphysical foundation supplied by the axiom is an incontrovertible, immutable absolute. Those who make the claim that order is proof of existence and, therefore, havw no justification for claiming and orderlyuniverse. It is the identification of the God of the universe that tells us the universe must be orderly, regardless of our shortcomings in our attempts to discover this order.

Conversely, however when that universe fails to correspond with the principles men formulate to describe, let alone explain it, this is also used as evidence of god. When the facts as we understand them turn out to be counterintuitive. This is regarded as a failure of reason, and evidence for a mystic god and a contradictory existence, with an ultimate reality that transcends human reason and is knowable only by mystic intuition. The god of a disorderly universe would certainly be the god of the mystic. In this context, it is important to reference the calamitous effect of failing to recognize the axiom as a self-proving assertion, not an assumption eternally waiting for extrinsic validation. As for the first principle of reason and the effort to achieve and advance a rational view of existence, it is claimed that while the law of identity may be true and useful, one cannot build anything from it. The philosophy of identism, the ideas expounded in this work, stands in opposition to such a claim and shows that a metaphysical philosophy and even a natural philosophy, can be constructed, primarily from this first principle, the law of identity, and a few basic and universally experienced perceptions.

This not to say that advanced science is not important in this regard, but much in contemporary science needs rigorous scrutiny, weighted against the axiom and the existence of identity. Indeed, the axiom is the ultimate key to, and foundation of an understanding of ultimate reality, for the universe was constructed from the thing to which the identity principle refers. The primacy of self-sameness gives one the most basic and general understanding of ultimate reality.

While there has been some recognition among logicians and philosophers, such as the neo-kantian philosopher, Afrikan Spir, that the identity principle must be ontological, the implications of the ontological status of the axiom have never been fully and properly formulated. Fundamentally, what identism offers to the intellect of humanity can be stated in a simple phrase " the existence of identity." This acknowledgment can resolve so many of the critically important historical questions of philosophy.

Most important, it is critical to the recognition of the axiom as absolute foundation and beginning point of knowledge. Metaphysically, it also can provide the basis for profound understanding of the universe and the very recognition that existence is a universe, that everything is connected.

Being is not a meaningless or indefinable concept, Martin Heidegger who tried to make an art form of cryptic contradiction, as well and others, amounts to a hopeless and foolish attempt to transform existence or being into some kind of part or characteristic.

An ironic tragedy in philosophy is found in the embracing of being-ness and nothingness. Such terms as non existence or unreal have no valid meaning in a metaphysical context. They only refer to perceptions or ideas that are out of synch with reality, such as what is implied by the statement "infinity is unreal."

Contrary to the foolish notion that the concept of existence is undefinable and invalid, the definition of existence is the simple and clear. Existence is everything. Existence is not a part or characterstic; existence is the totality of everything. The failure to subsume everything into this concept is to violate the necessary truth, everything exists. Only the totality of existence itself exists independent of other things. Everything else is connected and shares the same primary part.

THINGS, TIME, IDENTITY AND CHANGE

For anyone who is alive and sensing awareness is self-evident. This is true even if one assumed it is only some sort of self-awareness, trying to convinvce or deceive itself into thinking there must exist, an extrinsic world or universe. Presuming that perhaps, these sensations were all that really existed, would not change the fact of the existence of consciousness. Even assuming that all one knows is one's own existence does not deny the fact of the existence of consciousness, and it is self-evident that this consciousness has a multitude of components.

As consciousness is something that exists, it is also therefore, a self-evident fact that existence whatever it may constitute, is multifaceted. This self-evidence, again, would not be negated even if one held the Cartesian or solipsistic view that the external world cannot be proven because even one's internal world is multi-faceted as well. A multitude of sensations is self-evident proof of the fact that existence is multifaceted. Because reality is multi-faceted, one can mentally abstract parts from this totality. The thinking of and regarding parts as separate is methodological, pertaining to epistemology and one's efforts to understand. This mental abstraction of parts is a method of comprehension; it is epistemological, not metaphysics. Everything is in some regard connected to

all other things. This is proven by the primacy of identity, as all other things share the same primary part, self-sameness. Metaphysically speaking, parts are not, and cannot be wholly separate.

Apart from the totality of existence itself, entities, something existing completely by itself and independent, do not exist. The notion of entities is the mistaking of mental methodological functioning for something that is metaphysical, Parts are part of, but never apart from, existence. All things, apart from the totality of existence itself, are parts of the Greek philosopher, Heraclitus, who said "no man steps in the same river twice," the story is told that he lent a sum of money to someone. When he tried to collect on his debt, this man to whom he loaned the money, told him that he was not the same man to whom he lent the money. This story illustrates the impracticality of the philosopher's view but does little to address the apparent confkict between identity and change. Heraclitus thought everything was in constant charge and that change was the only constant.

Conversely, however, the principle of self-sameness would seem to be indicating things can, in truth, never change. The view cannot be reconciled with the incontrovertible truth that all A is A. Such a notion as things changing cannot be reconciled with the immutable fact of identity. The only sameness is sameness with self. The notion of things changing is to commit the metaphysical error of maintaining one thing as another.

A thing is itself; a thing is all that which constitutes its objective reality. The popular notion that a thing somehow is not the sum of its parts, not the totality of itself, and neither does the sum of its parts because that sum is the thing. If something is thought to be more than the totality of its parts, it is because all of its parts are not being accounted for. Also, one's grasp of a thing often may require that it is viewed contextually, comprehended in a broader context.

A thing can be anything and everything, as would be implied by such terms. When mentally abstracting things, there are few things that metaphysics imposes on us. It would be false to regard identity as divisible or reducible, and false to regard existence as a part. In addition, it is a contradiction to hold one thing as another. Sandeep Bisht as a child is not Sandeep Bisht as a fifty years old. They are not the same thing. It is perfectly appropriateto regard those two different parts as two parts of a larger thing. Sandeep Bisht. It would also be quite appropriate for Heraclitus to hold his debtor accountable for the money owed, and yes you can step in the same river twice. For the most part, nonetheless, what one regards as a thing is a matter of methodology, not metaphysics. While inconsistently maintained, the concept of things being expressed here is implicitly understood and utilized in human thought.

Things and parts represent an organization of facts and constitute the means by which one comprehends reality. It is common to regards humanity as things, but also thing. However, if one were to regard half of humankind, love, hate and a partridge in a pear-tree, as a thing, it would be no less a thing than humanity. This thing one has abstracted maybe less cognitively useful, but it is a thing, nonetheless. This thing abstracted here did serve the cognitive purpose of demonstrating that things can and must be regarded as anything and everything.

If what one regards as a thing is a part or parts of reality, if it exist, however divergent its parts, then it is a thing. It is obvious, but significant that different things can possess mutual parts. Things are all parts and all grouping of parts including the totality of existence. Things are everything that exists.

The number one is the mathematical equivalent of thing. The necessary truths that metaphysics imposes on what may be regarded as a thing, as well as its flexibility, also apply to the concept of one.

Much in mathematics is a matter of methodology, but with mathematics, as withh all human thought, it becomes detached from reality when the axiom is violated. As all human thought, mathematics is meaningless and groundless with axiom. There is truth in mathematics, as with all assertions, when self-sameness is declared, such as two and two are four. There are concepts in mathematics that are methodologically useful, but when taken literally or truly, would be metaphysically absurd and impossible, such as the notion of an actual infinity or zero.

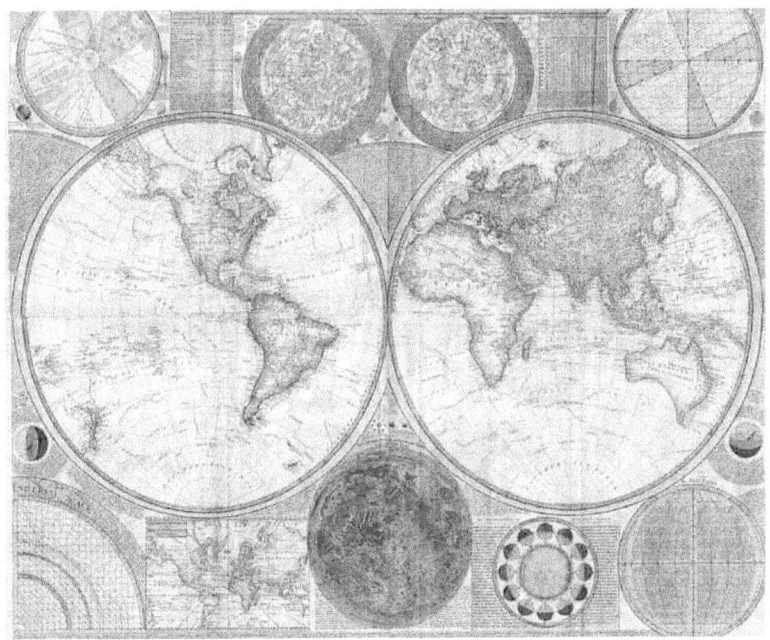

....

Eternalism represents a radical divergence from the commonsence assumption that has dominated the human understanding of reality and life. It is, nonetheless, an itellectual and spiritual ascension, that humanity must take, if they are to come closer to ultimate reality, understanding the supremacy of identity, and a metaphysical comprehension of its implications.

ACCORDING TO WESTERN CULTURE HOW GOD CREATED THIS UNIVERSE.

God created the universe according to eternal patterns in his mind and it is an expression of his thought, however incomplete an expression the cosmos may be. Erigena's pantheistic tendencies can be seen in his notion that God creates out of himself and "God is in all things." Creation is not in time but is eternal. In the process God used universals and made them particulars (e.g., humanity became individual persons). Immortality is the reverse process of particulars going back to universals. In Erigena's terms, division is the process of differentiating universals into particulars; analysis is the reverse, a return to unity and thus to God. These are not mere mental activities but mirror reality and God's relationship to the world. God is ultimately unknowable, being beyond all language and categories. Aristotle's predicates and categories cannot apply to God because they assume some type of substance. Nevertheless God can be described, albeit inadequately, using both positive and negative statements. Positive statements are only approximate but can be made more exact by adding negative statements. For example, it can be said that God is good (positive), but also that he is not good (negative) in that he is above goodness. These can be combined in the statement that he is "supergood." In spite of these approximations, God must be reached by mystical experience.

The universals are in things and have no existence apart from particulars. Objects are contingent in that they may or may not exist; they do not have to exist. Therefore there must be something that has to exist-that exists necessarily-to ground the existence of all other (contingent) things. This being is God. The world evolves by emanation, and matter is a phase of that process. The potential in matter is made actual, and over time God brings out its form. Thought is one emanation from God, and through it knowledge arises in humans. The actualized human intellect becomes an immortal substance.

The God that can be known in part from the universe is fundamentally different from it. Only God is identical to his essence, being neither more nor less than it. By contrast, a being such as Socrates is transcended by humanity because there are other people. On the other hand, Socrates has qualities ("accidents") that are not part of his essence; for example, he may be sitting. So unlike God, Socrates is both greater than and less than his essence. There is nothing that transcends God so nothing is greater than his essence. And there are no accidents in God because accidents are caused by something else (just as part of the cause of Socrates sitting is a chair).

God is not (completely) knowable because he is not material, whereas our knowledge is normally dependent on our senses. Furthermore, we normally know things by knowing their genus and species, yet God is unique and so cannot be known in that way. We can know something of God the negative way (via negativa) by removing limits, concluding for example, that God is unmoved, and unlimited by space. What we can know of God positively is neither exactly like our knowledge of temporal things (univocal) nor entirely different (equivocal). Rather, it is analogical, being in some ways the same and in other ways different. God knows x in a way that is both like and unlike the way in which Socrates knows x. God knows, but in a way that is, among other things, complete, immediate, and timeless.

That God created is evident (though not provable) because a material universe cannot emanate from an immaterial being. The universe exists to manifest God, who created the fullest possible range of beings because in them he can be revealed to the fullest extent. Beings range from angels, who are immaterial; to humans, who are material and immaterial; to animals, who are purely material (and both eat and move); to plants, to inanimate objects.

God as primary cause works through such created things as secondary causes. Nevertheless, creatures with a will remain free and responsible. God can also work apart from secondary causes in what we call miracles. Being good, God created the best possible world in the sense that it has the best kinds of things. Evil is a privation or lack of good and as such God did not cause it the way he causes other things. So we cannot ask why God brought about evil, but we can ask why he did not bring about more good. He did not bring about more good in order that he could be revealed through the greatest range of things, and as well, to allow for certain types of good (such as compassion, which can exist only where there is some suffering).

Aquinas and others grounded the scholastic synthesis of knowledge in the view that truth, morality, and God himself could be known by reason because the divine will itself is guided by reason. What is reasonable is therefore what is true and right. But John Duns Scotus (1265-1308) claimed that in humans and in God it is the will–not the intellect–that is primary. Evidence of this is that a being must will what to think about, thus something must act on the intellect; whereas nothing need act on the will. The view entails that there is no reason why God acts or wills as he does. This makes truth and morality essentially arbitrary and thereby unknowable through reason. God could have willed different moral standards. Scotus's view makes our knowledge of God a matter of revelation and faith, not of reason.

Another concept about God's will further destabilized the medieval world view. William of Ockham (1285-1347) held that omnipotence means God can do literally anything. Accordingly, a person could perceive something by sheer act of divine will, without the object being there at all. On his view, faith and reason can be contradictory. Ockham's "razor" sought to cut from explanations those entities that are unverifiable thereby making simpler explanations preferred. This was later used to cut out of world views such things as divine purposes, which had been central to explanations since the Greeks. Eventually, even concepts of a divine being would be optional–or even unnecessary–to explanations and world views.

The connection between reason and God was further undermined by Meister Eckhart's (1260-1327/28) view that God is "above being" and that human unity with the divine must be suprarational. Knowledge is a matter of proceeding from particulars to unity, beyond which is a unity with the divine surpassing all differences, "a silent desert." The divine being is therefore inexpressible. God knows all things in their unity, timelessly; but on our temporal level it makes sense to differentiate time as well as events.

Renaissance Thought

God moved out of the intellectual center of knowledge as faith was no longer grounded in reason and reason was no longer supervised by faith. The power of the church waned and society found inspiration in the classical world. Interest in this life and the world drove interest in science, which soon uncovered mathematically describable physical regularities. This development shaped the concept of God in a way that further undermined the Aristotelian world view, with its emphasis on such things as divine purpose. Regularities such as those discovered in Kepler's laws of planetary motion and Newton's laws implied a supreme engineer. Early in these developments, Giordano Bruno (1548-1600) emphasized God as immanent in the universe as an active principle, a trend in the conception of God that would increase along with the ever more detailed understanding of natural processes to be achieved in the scientific revolution.

WESTERN'S FINALITY AND NONLOCALITY

In 1935, Albert Einstein, Boris Podolsky and nathan Rosen published the "EPR papers" which described, using thought experiment, the nonlocality implied by quantum theory.

Nonlocality refers to the capacity of objects to instantaneously interact even as they are separated by large distances. immediate environment. Einstein referred to the principle of locality as an axiom, and it may have been a reasonably justified assumption. It is not, nonetheless, an axiom as the concept is defined in this writing. It is not necessary truth; It is not a self-proving assertion of identity.

Contradictory notion, or not, Einstein never accepted the existence of nonlocality and regarded it as evidence that quantum theory was false or at least incomplete. It was Einstein's abhorrence of nonlocality that supplied the motivation for his radical conception of gravitation, as he held that Newton's concept of gravitation violated the principle of locality.

In 1972, however, an actual experiment was conducted by John Clauser and Stuart Freedman. Despite Clauser's sympathy with Einstein's position, experiments confirmed the existence of this action at a distance. However, is this so-called nonlocality really so inexplicable? Space/Time is the realtive position of fundamental objects.

It exists in mutual causation with the fundamental objects it unites. It is not just out there, existing in and of itself. There may, or may not, be a direct space/time connection between two fundamental objects. If there is such a direct relative position between fundamental objects, then they are "entangled." This, according to finite geometry theory, is what is known as entanglement, If fundamental objects are "entangled." it is the finite, quantized, structure of space/time that provides the mechanism for what Einstein referred to as "spooky action from a distance." It can provide a simple, straightforward, non-mystical, explanation of so-called nonlocality. It is really, not "spooky" at all.

It is ultimately the assumption that two objects in direct relative position are disengaged that is irrational because it implies infinite divisibility of space/time. Since the finality explanation of interaction at a distance maintains that it is the specific constituent parts of quantized space/time that are in proximity with the fundamental objects interacting that causes the interaction, one can maintain that it converses Einstein's cherished principle of locality.

The finite structure of space/time provide the means of interaction and so-called spin. As spin down is weightier, it should have the greater space/time disparity. What is known as weight or gravitation is the result of the space/time disparities that result from fundamental objects in relative position concurring with finality. Quantum theory indicates how interconnected the universe is on a fundamental level; the finality thesis, the finite geometric structure of the universe, explains why this the case.

The God of reason

As identity is the only first cause the universe, therefore, must be determinstic. Essentially indeterminism is a claim that something other than identity can exist primary and causeless. As the primacy of identity proves, this violates the axiom all A is A. Nothing is exempt from causation; nothing is exempt from identity. Fundamentally, identity is causation.

Properly defined this does not mean humanity is without volition. Humankind has the ability to make choices, but this capacity is not indeterministic. There is really no contradiction between determinism and volition unless volition is transformed into a mystic notion, regarded to be exempt from causation, exempt from identity.

Determinism is certainly no reason to regard individuals as not responsible or accountable for their actions. To the contrary, it is precisely why must be held accountable. The failur to do this is to reward immorality at the expense of the ethical with destructive social consequences. Reality will hold such a foolish society responsible with unattractive results. Nature itself most certainly has and will hold individuals and humanity accountable for their choices and will pass the coldest judgment.

Volition, the capacity to make choices, is an immense factor in determining human survival and well-being. It is the human mind that must exercise volition, and it should be guided by facts and reason, but the fundamental basis for the choices one makes will ultimately fall back on basic convictions or beliefs. These could be notions derived from whim, religious fantasy, or some other irrational "belief system," or they could be ideas and principles fonuded on intellectual rigor. They could be factually and rationally justified convictions. It is only a first philosophy that can supply the necessary foundation for the later.

....

Philosophy is monumentally important to everyone, as it is essential to the proper function of the human mind in its relation to reality, and a recognition of the means by which humanity may live together in an ethical, beneficical, prosperous, and happy social environment. These significant examples certainly represent only a small sample of the diverse and important issues and questions that belong in the province of philosophy. Other disciplines can contribute too many issues that philosophy addresses, but should do so in a secondary manner. Even when philosophy is brought to bear on subjects that are not philosophy per sec, it performs a foundational function.

For the ones who are driven to question and pursue ultimate reality, it is both astonishing and tragic that so many live their lives in a state of object philosophy and acquiring any kind of metaphysicalperspective. Others dogmatically cling to the demonstrably false mystic paradigms provided by religion and statist propaganda. Human efforts in the realm of philosophy historically and the state of academic philosophy today are certainly, to great degree, responsible for the general philosophical vaccum that exists among non-academics. This alone does not explain the apathy that exists in the most people, for such important matters that, as Socrates said, to ignore is to make one's life scarcely worth living.

The practical necessity of philosophy, the important of commonsennse rationality being ascended to metaphysical, are willing to surrender their yearning for metaphysical knowledge, to fantasy and delusion. Usually, when philosophy does emerge from outside the academic establishment, it is every bit as horrendous as what the establishment offers. In many cases, it is much worse. Be it, intentional or not, irrational ideas coming from outside the established mainstream serve as "strawman" arguments for the establishment.

....

First philosophy, by its very nature, should be universally relatable. True first philosophy must being with the self-evident and self-proving axiom of identity, and the most basic rudimentary facts that should be universally acceptable, as self-evident. Reading a philosophical work should not be like reading a technology journal, filled with concepts that reference specialized knowledge with which many are unfamiliar. True first philosophy is fundamental and should be comprehensible to the intelligent layperson.

The fundamentals of identism are easy to understand, and to one who is intellectual honest, easily regarded as provable. This basic principle that it is initiated, and concepts that follow, nonetheless, lead to ideas and stunnig revelations that are deep and in some regards, counterintutive, constituting a very different view of existence, the world and the ultimate nature of one's life. In this regard, it may emerge as a metaphysical view that is for many, challenging to conprehend fully.

A true and well comprehended metaphysical philosophy and reason grounded in an unshakable recognition of knowledge must prevail over mysticism, debilitating relativist dogma and pseudoscience parading as objective research before mankind can realize his true intellectual potential. Metaphorically speaking, first and foremost, humanity must discover the God of reason.

Identism is a philosophy centered on the supremacy of the axiom and the primacy of identity. In a meaningful, but only metaphoric sense, identity is the God of the philosophy of identism. There are interesting parallels between the Identist conception of identity and the mystic notion of god. As the creator gods of the faithful are generally envisioned, identity is the primary thing from which all other contingent things derive their existence. The law that asserts identity, the principle of self-sameness, is the supreme law of existence. This mirrors the notion that the creator gods of religion are the ultimate authority.

When a believer is called upon to justify their mystical belief system, they will usually fall back on the words of some ancient text that is claimed to be divine. In parallel to this, it is the axiom formulated with words that describes, asserts and proves the existence of identity, the metaphoric God of reason. Religion claims great harm in failling to acknowledge the existence of their gods, but it is the failure to acknowledge the existence of identity, God of reason, which has created great harm to human intellectual potency.

....

Religions claim their god is the path to salvation, but it is the God of Reason, that stands as the only hope of saving humanity from these destructive and debilitating gods of the mystic. Much as the gods of faith are jealous gods, The Gods of reason must be worshiped uncompromisingly to constitute an incontrovertible claim to knowledge.

In addition to parallel between the God of reason and the mystic gods of faith, there are diametric opposites. First and foremost, ultimately the belief in the gods of faith is a belief in non-identity. The gods of the mystic as they are generally defined, when their believers can be imposed upon for a definition, are contradictory, non-identity, which cannot be reconciled with the fact of the existence of self-sameness and can only be accepted on faith, on self-deception. The embracing of identity, contrariwise, constitutes a rejection of faith, a rejection of belief. While the gods of faith, on self-deception. The embracing of identity, contrariwise, constitutes a rejection of faith, a rejection of belief. While the gods of faith sabotage any honest claim to knowledge, the God of reason and knowledge are discovered together. The God of reason is the certain immutable ground and starting point of knowledge.

Faith is a delusion, falsely ascended to the status of vistue. The gods of faith are fought for, with the weapons of murder, in the horrendous battlefields of the holy wars of human carnage and sacrifice. Aside from psychological ploys aimed at human self-deception, this is the only means by which such gods can be established. Conversely, reason soundly founded on the absolute is the only weapon required by the worshipers of identity, the God of reason.

For the worshiphers of the God of reason, there is to be found an element of revenge and yet, transcendence and forgiveness, for the little lying blasphermers can oly be exactly what that which they deny has made them. For many, their whole intellectual life, and consequent efforts are wasted in a pathetic attempt to evade and deny the absolute. What they embrace, however the imaginary non-identity, in all its embellished and gilded variations of self-deception, is destroying the world.

Therefore, the liars must be confronted and human knowledge must prevail.

To embrace the gods of faith, non-identity, is to abdicate any honest claim to knowledge. Conversely the embracing of the God of reason identity, is the only means by which one acquires knowledge. Embracing identity is to acquire the power of knowledge. It is a power that has escaped humanity with catastrophic results.

Unlike the mystic gods of faith, the God of reason does not ask for victims or sacrifice, does not ask for blood. It does not demand that its proponents drag victims to the top of pyramids so their hearts can be ripped from their bodies. It makes no demand that innocent children must beg for forgiveness. While it is ultimate root of the grief, suffering and misery in the world, as it is the cause of everything, it is the one god that perphaps we can forgive and accept.

This is the spirituality the God of reason offers. It is the only spirituality worth possessing because the price is much to high for the fraudulent "spirituality" offered by the mystic, for it demands the surrender of the mind, of knowledge. It is a sellout to the worst kind of insidious evil. The price that has been paid for this false delusion of happiness is an incalculable failure, suffering, pain and guilt. Religions cannot conjure the gods or heavens they hallucinate, but they certainly can manufacrture the mystics hell on earth.

The embracing of the God of reason does not contradict real human needs. It does not stand in antagonism to the very values required for human survival. It is spirtuality grounded in the very source of human power, survival and well-being. Conversely, the gods of faith and their fraudulent "spirituality" and "religious experience" are the principle threat to the existence of humankind.

Through history, humanity has been dominated by the gods of faith and the monstrous delusion they bring forth, greatest and most hideous of which, is the state. The emergence of humankind, nonetheless, has been the result of the latent effect of the god of reason, of the intellectual power it brings to humanity. The incompatibility of human progress, and the primitive monster that has been dragged along, is reaching a monumental crisis and potential catastrophe beyond anything in human history. When Patrick Henry stated, "give me liberty or give me death," it was, at the time, declaration of defiance. Today, however those words convey a grim statement of fact. This is the alternative humanity may well be facing.

Groundless and floating, human advance itself becomes a source of prodigious danger, concern and crisis. In the face of human achievement, mysticism and statism become more and more, deadly anachronisms. Human reason ha screated technology of incredible power and potential, while human irrationality and propensity for self-deception has placed it in the hands of monstrous criminals.

Truth it has been said "needs no protection," but lies have have always enjoyed the protection of the state, as the state has enjoyed the protection of lies. It is, therefore, the state from which the truth must be liberated and protected. Statism has always been poisonous to scientific objectivity and honest metaphysical inquiry. Politicized and intellectually corrupted science at its worst can replacereligious mysticism as the enablers of the criminality of the State. Such "science" has deteriorated to the level of mysticism. This certainly, is also true of tainted, dishonest or wrongheaded philosophy.

Regardless of whether it is called science, religion or philosophy, lies and self-deception are vices that enable and perpetuate the state. From the most primitive social orders forward, religion, mysticism and pseudoscience have formed alliances with the violent oppressors and their fraudulent claim to authority that are noe known as government and the state. Only reason founded on identity truly stands in antagonism to this evil alliance.

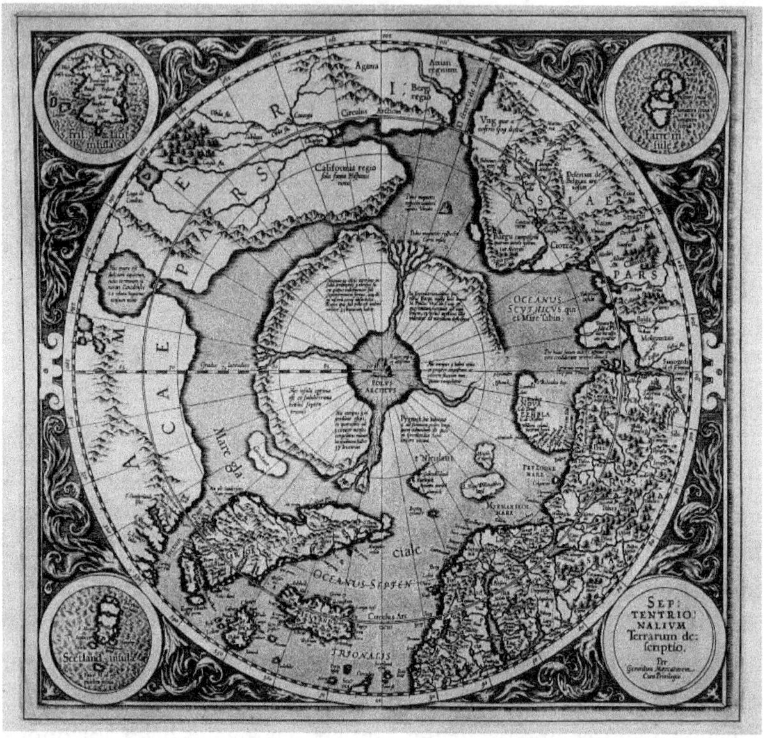
SEP: TENTRIO: NALIUM Terrarum de: scriptio.
California regio
OCEANUS SCYTHICUS
Marc gla ciale
OCEANUS SEPTEN TRIONALIS
Circulus Arcticus
Scetland insula

....

True spirituality is the product of knowledge of reason. Spirituality clearly lies in the realm of the human capacity for truth and must be founded onthe recognition of identity. To embrace contradiction is to create disharmony in the spirit, conflict in one's mind. True spirituality cannot be founded on mysticism; it cannot be derived from a premise that would destroy the very concept of truth.

When the mystic peddles something called spirituality, it is really its antithesis that he offers, self-deception, delusion, and mendacity. It is blindness, ignorance and above all, lies; it is contradiction that divorces one's mind from existence the world, and ourselves. Spirituality, like knowledge and truth, can only be founded on the law of identity. Of the grand achievements that the principle of self-sameness makes possible, and can onlymake possible, spirituality most certainly is one of them.

There is much more at stake, nevertheless, than mental tranquility. It is a quest to make a claim to the very means of human survival; it is claim of the power and competency of the human mind. This serves as the only true basis of one's claim to natural rights, to the natural freedom humanity possesses unobstructed by criminality. To refer to such freedoms as right is to make the moral claim that it is right that humans possess such freedom. it is a moral claim that is rendered mute if humanity is to degenerate into vacuous, mindless, sacrificial animals.

The recognition of natural rights is an intellectual achievement that cannot be grounded in faith or mysticism. It can only come through knowledge, which is only achieved by embracing the axiom and existence of identity.

Faith never has, and never will, serve as a successful basis for the advocacy of freedom because it extinguishes the very thing that makes freedom an indispensable requirement for human survival and well-being. The blind, ignorant mystic has abdicated his claim to freedom and natural rights.

····

Reason must find its God. If humanity truly strives for the attainment of a meaningful "salvation" there is but one God that can provide this value, if only humanity can discover the God of reason, that which created the universe. It is not human reason that is falling. It is humanity falling reason, the best within them, by refusing to grasp and embrace the God of reason. All A is A!

GLOSSARY OF TERMS

The axiom is self-proving assertion of identity such as "all A is A." Identity is existent thing assertedby axiom. Self-sameness is identity recognized as ontological and acknowledged as a part of all else. Specificity is identity acknowledged as a part of all other parts. Finality is identity recognized as a part of quanitty, the self-sameness of numbers. Quantity is existence acknowledged as multi-faceted. A given quantity is a given thing. In this regard quantities are things.

Existenceis everything. The failure to knowledge this is to violate the necessary truth; existence, if it truly is, exists. Existence cannot be added to or subtracted from; existence is unchanging. Reality is existence. The term is often used in the recognition that existence may, or may not be, correctly perceived or comprehended.

The universe is existence understood by the primacy of self-sameness to be united by a mutual part, identity. Everything is connected. Knowledge in the absolute sense is truth held with justifiable certainty. However sophisticated one's information about existence, without this ground of absolute certainty, one's very grip on reality is tenuous. This is stunningly and disturbingly illustrated ny the present state of humanity and the condition of the collective human mind. Knowledge can only be achieved by recognition of the immutability of the axiom and the existence of identity.

CONTACT AUTHOR

- Email: Sandeepbisht256@gmail.com
- Contact no.: +919319908754
- Insta: https://www.instagram.com/sandy41bisht/?hl=en

www.ingramcontent.com/pod-product-compliance
Lightning Source LLC
Chambersburg PA
CBHW070813220526
45466CB00002B/652